PAST & PRESENT

BOULDER CITY

OPPOSITE: Prior to cold weather, the canyon workers were transported to the Boulder Dam site from Boulder City in buses. The dormitories were for single men, and the operation became routine. Every day, the transport trucks unloaded exhausted men, and then returned to the dam with a fresh cargo of workers. Day shift started at 7:00 a.m. and knocked off at 3:00 pm. The swing shift (pictured) was from 3:00 p.m. to 11:00 p.m. Graveyard shift was from 11:00 p.m. to 7:00 a.m. They battled the Colorado River 24 hours a day. (Boulder City Museum and Historical Association Collections.)

PAST & PRESENT

BOULDER CITY

Tiane Marie

To my father, I dedicate this, my first book, to you. You instilled a love of history, geography, and home in me for which I am very grateful.

Copyright © 2021 by Tiane Marie
ISBN 978-1-4671-0750-1

Library of Congress Control Number: 2021943537

Published by Arcadia Publishing
Charleston, South Carolina

Printed in the United States of America

For all general information, please contact Arcadia Publishing:
Telephone 843-853-2070
Fax 843-853-0044
E-mail sales@arcadiapublishing.com
For customer service and orders:
Toll-Free 1-888-313-2665

Visit us on the Internet at www.arcadiapublishing.com

ON THE FRONT COVER: Denver Street is a gateway back road as one heads to Lake Mead and the Hoover Dam. The home pictured belonged to Charles and Mary Ida Cady, pioneers of Boulder City. Cady spent much of his life serving in the US Bureau of Consular Affairs and was associated with the Benguet Consolidated gold mine in the Philippine Islands due to his engagement in the gold-mining business. (Past image, Boulder City Museum and Historical Association Collections; present image, author's collection.)

ON THE BACK COVER: Pictured is Ida Katherine Browder Kelley, daughter of Ida M. Browder, on the running board on site at Browder's Lunch in 1932. The Browder building was the first building to be constructed in what became known as the "free enterprise" section of the government town, and it marked the boundary between the Six Companies area and independent business sites. (Boulder City Museum and Historical Association Collections.)

Contents

Acknowledgments — vii
Introduction — ix

1. Roads and Streets — 11
2. Restaurants and Businesses — 31
3. Housing — 61

Acknowledgments

First, I want to thank my daughter, Leora Rose, for being my biggest cheerleader, for without her constant cheering, the completion of this book would not have been possible. You are my world, and any creative idea without you is just fiction—I love you.

Second, I give incredible thanks to the Boulder City Museum and Historical Association for its broad range of "past" images and information. The time and detail it took to weed through the photographs and choose only a small group of images and information was hard enough. However, without such a selection, this book would have taken longer to be birthed. All "present" images are courtesy of the author and a friend who wishes to be anonymous.

I also owe a special debt of gratitude and thanks to my parents for their constant help and patience with me during this process.

I have a few fantastic resources who provided guidance and support during this process and wish to remain anonymous, and I would like to express my greatest appreciation to you.

Introduction

Located in the extreme end of the southern tip of the state of Nevada—only a few miles from Arizona, across the Colorado River, and about 40 miles by plane from the boundary of the state of California—is Boulder City. This town was designed in what is known as the harsh desert environment, and its sole reason for existence was to house workers that were contracted to build Boulder Dam. When the proposal was made to build a dam to harness the Colorado River at Boulder Canyon, the nearest human habitation of any size was 30 miles away in Las Vegas. Once the precise site for the dam had been declared by the US Bureau of Reclamation in 1930, single men and men with families began settling in tents along the Colorado River in an area known as Ragtown.

On March 4, 1931, it was announced in Denver that Six Companies, Inc., had made the low bid for construction of the dam. With the flood of people showing up in Ragtown on a daily basis, the government realized it needed to quickly build a temporary town to house the workers. By March 24, 1931, the first eight houses in Boulder City had been completed, and an additional eight were under construction. This town was allowed a path of hope due to the Boulder Canyon Project Act, giving the people a pride that flourished from rags into a community. The town had a population of 2,000 after all the features construction ended on Boulder Dam in 1936. In 1940, the town was estimated to have approximately 2,738 residents. The influence of such an influx was due to the activities of the federal government in Boulder City, security precautions, and the attractive features of the town, which was filled with professional office workers, skilled laborers, magnesium plant/Hoover Dam workers, and essential service businessmen.

The people are the heart and soul of Boulder City. Boulderites were churchgoing, organization-forming, proud, and community-minded folks. Some of those who were original 31ers (meaning that they arrived in Boulder City in 1931) remained a part of the community, while some moved on to the next job. Their lives were probably about what most people would have in mind when they think of a small town between its inception in 1930 and becoming an incorporated town in 1960—from a makeshift shantytown consisting of tents to a district with banks, movie theaters, newspapers, a department store, a hotel, churches, a library, etc. There was a quiet dignity in being the operating headquarters of a town, and the overall character was one of a mixed-government area, but it became a community that has provided some of the richest historical quality of life in America.

CHAPTER 1

ROADS AND STREETS

This image shows the view from Railroad Pass, which opened in 1931 just outside the border that surrounded the Boulder Canyon Project's federal reservation. Due to the strict reservation rules, Railroad Pass was in the perfect location just over the border to the west. Dam workers stopped in after a week on the job and stayed to enjoy an illegal drink, gamble on roulette, or listen to the five-piece band. To get in, one had to know the password—gaiety.

This view of Boulder City is looking south down Nevada Way from Avenue B/Arizona Street in 1939. Proprietors of the Green-Hut Café (pictured at right), C.D. Newlin and Frank Aster, opened the establishment in April 1932. It was one of the town's largest and finest cafés, located near the Troy Laundry building. It was furnished with the most modern equipment of its time, which came from Los Angeles.

The hub of the community's businesses was at the intersection of Nevada Way and Avenue B. There were rapid strides to create a business district, especially with the town being only 13 months old at this point. The six new businesses in 1932 were Manix and Vaughn (arches at right), a general store, Boulder City Drug Ltd., Green-Hut Café, Miller Brothers' meat market, Dr. McItosh and Dr. Parks (dentists), and Nelle McDougal's Powder Puff beauty shop.

ROADS AND STREETS

When Saco DeBoer designed the town, he was mindful of how miserable summer heat could be and advised stores and shops in the district to be fronted with arches and arcades over the sidewalks to protect visitors. However, after World War II ended, these characteristics were not considered to be in style anymore. Some decided to demolish their arches, while others stayed true to the practical and aesthetic value of them.

ROADS AND STREETS

Below, behind Walter Vaughn of Manix and Vaughn and Charlie Herring, in the distance on the other side of the road, is the structure (with white arches) that later became Glinski Boulder City Motors in downtown Boulder City. Bogdan "Bob" Glinski ran three southern Nevada auto dealerships between 1950 and the 1980s and became a household name for hosting late-night movies. In the 1960s, he unsuccessfully ran for county treasurer and the Nevada Assembly but did serve on the Democratic Central Committee.

Roads and Streets

On October 20, 1933, this was the intersection of Avenue B and Nevada Way, looking south. In 1944, Robert "Bob" Broadbent had a fresh pharmacy degree and decided to move to the federally operated town to start a drugstore (right) located within the booming business district. Even though he had increasing political and government duties, serving as mayor for one term and as a councilman for eight years, he ran Boulder City's principal drug dispensary for 25 years.

In 1935, at Nevada Way and Avenue B and Arizona Street stood a directory where tourists could find a map of the historic district with directions to food, shops, and businesses. Only three of Boulder City's restaurants were completely unionized—the Green-Hut Café, the Reservation Grill, and Grand Café. The social center of town was patronized by scruffy workers from the dam project, Hollywood celebrities, and visiting politicians.

Roads and Streets

The businesses pictured here in 1937 are, from left to right, Browder's Café (in the B.B. Thompson Building), the Nava-Hopi Indian Store, the Green-Hut Café, and the barbershop, which was owned by Jack Higgins, a Nevada state assemblyman from Boulder City. After the new Atkison Building was constructed by Horace Shidler & Son on Arizona Street near Nevada Way in 1946, the Nava-Hopi Indian Store was one of the businesses that moved into it.

The intersection of Avenue B and Nevada Way is pictured here in 1933. The Texaco station (right) was sold in 1940 to Henry J. Cottle of Augusta, Montana, a friend of J.C. Manix. Cottle had operated a Texaco station in Augusta for 14 years. A.A. Scotty, who sold the service station to Cottle, remained in charge of the Texaco until January 1941, by which time Cottle had fully relocated his family to Boulder City.

ROADS AND STREETS

This view looking east shows Arizona Street from Nevada Way at what locals called the "heart" of the town. Administration for the school (at distant left) was the sole responsibility of city manager Sims Ely. Those employed were chosen by Ely himself and paid by Six Companies and Babcock and Wilcox, which was the second-largest contractor on the dam project. In fall of 1933, the government turned over educational control to the State of Nevada.

Workers from the New Mexico Construction Company are shown installing curbing along Birch Street in 1931. The company received a $300,000 contract that covered the installation of water distribution and sanitary sewerage systems, including the grading of streets and alleys, automobile parking, and the laying of concrete curbs, sidewalks, and gutters. Its biggest operations were trenching pipeline for more than 23 miles and laying 90,000 square yards of asphaltic concrete pavement.

ROADS AND STREETS

This view of Boulder City is looking south from the top of Wilbur Square toward Escalante Plaza (center), also known as City Hall Park, and California Avenue in 1932. The central plaza's name is linked to the first European to cross the Great Basin, Silvestre Escalante. The history of western exploration and the Colorado River's geology were used to name the streets and plazas within the town in an effort to expand folks' knowledge of the area.

ROADS AND STREETS

A tourist is shown checking a sign that shows the direction and distance from Boulder City to various locations in 1936. During 1935, it was estimated a total of 365,000 persons visited the dam and the lake. This sparked a general plan under which the National Park Service and Bureau of Reclamation would cooperate in their control of Lake Mead for recreational purposes.

ROADS AND STREETS

23

This view is looking south down Utah Street with the Six Companies Hospital addition under construction in the distance. Sims Ely, city manager; Dr. Wales Haas, superintendent of the hospital operated by Six Companies; and sanitary engineer D.M. Forester were appointed to the board of health due to deaths from a disease—classified as influenza—that reached a total of 41 in 1932. Aside from the epidemic, the health of the town as a whole was excellent for a population of 5,000.

In November 1930, during the beginning stages of planning Boulder City, the Department of the Interior decreed that no car would ever park at the curb in town. There would be areas provided and conveniently arranged throughout the town to take care of parking demands in a modern age. A 1950 report of a survey made under the direction of the Bureau of Reclamation stated: "The Transcontinental highway on which the town will be built would wind through parks and residential areas rather than through the business district."

GOVERNMENT ADMINISTRATION BLDG. AND PARK, BOULDER CITY, NEVADA
BUREAU OF RECLAMATION PHOTO

Prior to 1932, the Bureau of Reclamation leased offices in the Beckley Building in Las Vegas as it waited for the principal buildings to be constructed near Boulder City. In August 1931, the construction contract was awarded to B.O. Siegfus of Salt Lake City for $46,253, with the government furnishing the materials. The administration building (pictured) was constructed on the great hill north of Boulder, and it could be seen from anywhere.

ROADS AND STREETS

Rudy and Violet LaCroix lived along Utah Street, which is shown in this view looking north toward the Bureau of Reclamation administration building in 1932. Rudy's law enforcement career brought him and his family to the area in 1941; he served on the Boulder City Ranger Force. The LaCroixes soon became popular participants in civic, fraternal, and social affairs. Rudy belonged to the American Legion, Elks, Independent Order of Odd Fellows, Patriarch Militants, Knights of Pythias, and the Retired Civil Employees Association.

Nevada Way at New Mexico Street and Birch and Cherry Streets is pictured here in 1932. In 1943, Bureau of Power and Light superintendent E.P. Bryant announced the bureau was building 20 new dwelling units between Birch and Cherry Streets for any employees who were married. The units would be in the nature of temporary cabins and would be completely landscaped with trees and shrubs. The contract was given to Knap California Inc. for $38,166.03.

The El Rancho Boulder Motel owners were Albert and Oleta Franklin. Albert came to Boulder City in 1932 as an employee of Six Companies to work on the dam. He then entered the construction field, building the Franklin and Lucky Motels in Las Vegas and the El Rancho Boulder Motel in 1953. The El Rancho was known for having the first swimming pool in town. Albert served as a member of the original charter committee and was one of the first council members after the incorporation of Boulder City.

ROADS AND STREETS

The Ash Street trench for a sewer line was dug in 1931 for homes that were originally built for Six Companies engineers and superintendents and acquired by the Los Angeles Bureau of Power and Light in 1936. Pete Cole, pipe fitter and plumber, helped make this happen. The 10-inch line, which carried water from the river up the hills to the water tanks that overlooked the growing town, was one of his babies. Cole also helped lay a 14-inch line.

CHAPTER 2

Restaurants and Businesses

Pictured here are the Six Companies recreation hall (left), Six Companies dormitory with chimney (center background), and Boulder City Company Stores (right). In their off hours, many workers went to the rec hall for relaxation, the pool tables, the soda fountains, the cigar and candy machines, and—after 1933—an occasional beer with a 3.2 alcohol percentage.

An interesting debate among locals is whether the Six Companies' temporary buildings are shown on what is believed to be the site of the present-day Boulder City Airport on April 18, 1931. When Six Companies first came to town, it built small cabins to be used as bunkhouses for employees. These were constructed before any other housing or administration buildings were planned. One of the buildings was used as an isolation ward when the hospital was constructed.

The Nickell Building housed the Chicken Shack Restaurant and Desert Sands Pottery. While Terrell Evans was employed by the Bureau of Mines, he set up a pottery at his home, where he created on weekends and in his spare time. This allowed his Desert Sands Pottery to be sold in southern Nevada gift shops. Later, Evans went from a work shack to a modern building to be within the business district and supply the high demand for pottery in Boulder City.

RESTAURANTS AND BUSINESSES   33

Pictured here in 1932 are the Manix & Vaughn department store (center) and the Miller Brothers meat market (center left). Miller Brothers was in the alley—known as T-Bone Alley—between Manix & Vaughn and Delmar's drugstore (at far left). J.C. Manix arrived in town from Augusta, Montana, with the vision of opening a business. He received one of the first business permits awarded to an individual. Manix and his partner, W.D. Vaughn, opened the department store in 1931. It contained every item one would ever need, from groceries to furniture and even apparel.

RESTAURANTS AND BUSINESSES

The corner of Arizona Street and Nevada Way is where the Shell Automobile Service Station was located. This picture was taken during the grand opening on March 6, 1942. Robert "Bob" Conners and John Keisling took over the operation in 1952. Conners came to town in 1942, during the Second World War, and after his service, he moved to Boulder City with his family in 1946. Keisling came to town in 1945 and worked as an employee at the Shell until he was made manager in 1951; he became a partner in 1952.

RESTAURANTS AND BUSINESSES

Located in what is now Sun Dial Park, the Boulder City Company Store—once the biggest department store in Nevada—occupied an entire block in Boulder City. The store was equipped and provided for the needs of dam construction workers and their families. It also had its own unique money system, and workers could take credit there as well. After the construction of the dam was completed, the company store was demolished.

The first location of Roy Fairbanks' Men's Clothing is shown here on December 1, 1931. The business moved a few buildings down and to the left in 1932. Fairbanks was a pioneer resident of Boulder City during the dam-construction era. Not only did he build the clothing store, Fairbanks also constructed the Emporium. After selling the men's store to R.L. and R.W. Georgeson in 1939, Fairbanks traveled extensively around the nation, only staying in town between his trips.

RESTAURANTS AND BUSINESSES

After the repeal of Prohibition, the first 3.2 beer and 4 percent wine was allowed in Boulder City and was served here at Laubach's Recreation Tavern. It was not until 1969, after the city was incorporated, that the first bottle of liquor was sold in town. This building remains a tavern and is now known as The Backstop. One tradition is still in place—the tavern offers a free beer to everyone any day the sun does not shine in Boulder City.

The Boulder City Elementary School is shown in a view from North Escalante Plaza in 1933. It was built by the Bureau of Reclamation and operated with Six Companies funds. The building was designed with a distinct brick and was not finished until September 1932. Six Companies contacted the school district in Las Vegas to find out what books to stock in its store. Residents had to purchase these books for the children, because the school did not have any equipment.

RESTAURANTS AND BUSINESSES

39

Boulder City's Municipal Building was completed in 1932 and housed the post office, a courtroom, the police department, a jail cell, and a kindergarten room. Because the town was a federal reservation, there was no police force; the area was maintained by a US marshal and government rangers. The rangers mainly ensured the reservation was clear of unauthorized people and enforced the liquor laws. The most "crime" the town saw involved 48 people who broke Prohibition laws and authorities removing more than 1,000 "undesirables" from the town.

RESTAURANTS AND BUSINESSES

Ida Browder, Boulder City's first businesswoman, was known as "lady of firsts." After becoming a widow, she moved to Boulder City in 1931 to support her two children. After quickly obtaining permission to build a diner, she created the first permanent commercial establishment. She was actively involved in the community, operated an informal savings bank for dam workers, and served affordable meals. She was like a second mother to many of the young men who worked on the dam.

Restaurants and Businesses

St. Christopher's Episcopal Church is located at the northwest corner of Arizona and Utah Streets. The historic church was completed in 1932 and is one of Boulder City's two original churches that still exist today. Because the town was considered a reservation, St. Christopher's was built with Episcopal missionary funding. The first services were held by Rev. Arthur Keane. It stayed a quiet congregation due to its conservative views until the 1970s, when controversy arose.

The historic Boulder Dam Hotel operated under a government permit beginning on October 20, 1933. Constructed with the vision of Jim Webb in order to accommodate a growing tourist industry for the building of the dam, the hotel gained its popularity thanks to its fine hospitality and a richly appointed interior. Today, it is the only building within the Boulder City Historic District that is listed in the National Register of Historic Places, and it still provides an authentic historical experience for tourists. It is also home to the Boulder City/Hoover Dam Museum.

RESTAURANTS AND BUSINESSES

43

After Earl Brothers completed the construction of this building in 1941, the visitors bureau moved from its location at the theater into this building, as did the bus terminal. Later, the building housed a drugstore and bank. The visitors bureau remained in this location until 1964, when a fire broke out inside the theatre. This is one of many structures in the historic district with a unique International construction style.

The Boulder Theatre, located at 1225 Arizona Street, was built in 1932 by the modern-day pioneer Earl Brothers. The cooled theater offered free movies to dam workers 24 hours per day as a relief from the extreme heat of the desert. As the theater grew in popularity, Brothers established the Boulder Dam Service Bureau in the upstairs of the theater, where tourists could purchase souvenirs. In 1939, Uptown Hardware (1229 Arizona Street) was added.

Restaurants and Businesses

The new high school was located east of Boulder City Elementary School on Arizona Street and opened on January 2, 1941. The building included a gymnasium/auditorium, and the first class graduated in 1942. Prior to construction of this building, Boulder City high school students were bused to Las Vegas and graduated from Las Vegas High School. An addition was completed to the rear of the building in 1991. This building currently houses the Boulder City Parks and Recreation Department.

In March 1932, the first load of lumber was dumped on the site of Boulder City Lodge No. 37. The structure was completely built by the hands of lodge members, from the concrete block to the roof trusses. The cost was approximately $2,700, and ground rent—due to the Boulder City reservation—was $1 per year. The space is currently shared by the Order of the Eastern Star, Rainbow Girls, Chapter, Council and Commandery of York Rite Masonry, and the Boulder City Shrine Club.

RESTAURANTS AND BUSINESSES

47

In early 1931, a small fruit stand was owned by Albert "Al" Stubbs and a partner. As it flourished into a larger market, it became what was known as Central Market on Wyoming Street (pictured). Central Market moved into its new location on Arizona Street in 1948. Stubbs was a pioneer whose descendants represented Boulder City, including Bishop Leonard Stubbs, chief of building, grounds, and communications for the Bureau of Reclamation.

In December 1950, William Belknap took this photograph of the Boulder City Belknap Photographic Service building. Belknap, Mark Swain, and Cliff Segerblom opened the center. The three talented photographers had traveled around the world but decided that Boulder City was their place to grow. The shop was a stopping-off point for tourists and locals in Boulder City. Segerblom designed the unique building that still stands today.

RESTAURANTS AND BUSINESSES

Pictured in front of the Bureau of Reclamation building in 1939 is Major Merrick with the Civilian Conservation Corps. When it was completed in January 1932, this building became the Bureau of Reclamation's most substantial and imposing project headquarters. The first floor housed the offices of the chief construction engineer, chief clerk, and other staff members during the construction of the dam. The second floor was designed to accommodate the district counsel, visiting engineers, and consultation/drafting rooms.

50   RESTAURANTS AND BUSINESSES

The engineers' dormitory and annex No. 1 are shown here in March 1932. The dormitory was one of the buildings Saco Rienk DeBoer had to incorporate, as it was required (by the Secretary of the Interior) to be part of the bureau's headquarters. DeBoer was hired in 1939 to design the town and spent the better part of the decade doing so. Developing two distinct residential areas separated by a forested beltway is how he described the difficulty of designing what was intended to be a temporary community to house workers and government officials.

Restaurants and Businesses

Six Companies Hospital served as the town's only hospital from 1931 through 1974. It was then used as a religious retreat center until 2015, when the city council voted to have the building demolished. In its prime, the hospital served the dam's construction workers and their families. It was a 60-bed facility headed by Dr. R.O. Schofield, chief surgeon, and staffed by six doctors, nine full-time nurses, two full-time orderlies, and several support personnel.

The water filtration plant was built in 1931 as part of the water supply system. The original system, created under the auspices of the Bureau of Reclamation, included a pipeline from the dam, a pumping plant, a filter plant, and associated storage. A supplemental system with an additional pumping plant and pipeline (parallel to the other pipeline) was added in 1949. The systems supplied treated water to Boulder City residents, who relied solely on the Colorado River, since there is no potential use of local groundwater.

Six Companies machine shop and warehouse, also known as the supply center, are pictured in 1932. These buildings were served by railroad spurs. A spur line off the main Union Pacific tracks was used for the railroad to haul equipment and construction supplies through Las Vegas for the Boulder Canyon Project. As the last installation for the dam was finished, the need for such a line diminished.

The first grocery store/café in Boulder City was located at the bottom of Colorado Street on the truck route to the dam. Pictured here on April 18, 1931, are (left to right) W.F. Shields, owner; Harry Buchanan, cook; and a Mrs. Shields. The Shields brothers had owned a store some years prior in Hesse's Camp on the Colorado River, near the mouth of the Las Vegas Wash.

RESTAURANTS AND BUSINESSES

Hunger Cure was located on lower Colorado Street, west of Birch, facing the truck route. It had a sheet-metal counter and apple boxes for seats. It was open for months when Jack Shields, the brother of W.F. Shields, who owned the first grocery, bought the restaurant. However, after a few months of operation, it was shut down due to health authorities citing Shields for serving wild burro meat—a humble beginning for the restaurant boom of Boulder City.

Restaurants and Businesses

The Union Pacific Railroad completed a branch line to the area along the entrance to the town in early February 1931, making possible the construction of Boulder City. At center are the Union Pacific Railroad–constructed oil tracks used to serve Standard Oil Company of California and Union Oil Company of Nevada at Boulder City. On this block (103), these buildings were used for the companies' storage tanks, warehouses, and offices.

Restaurants and Businesses

Prior to the expansion shown in this 1941 image, Grace Community Church began as a small, Protestant, interdenominational, religious church and was supported, in 1932, through the combined efforts of seven denominations. Their thought process was to "let there be one Protestant church cooperating, not a dozen competing" in the booming town. The church broke ground—and the first service was held—in 1933. Rev. Thomas Stevenson, from the California Presbyterian Church in Burbank, was the first minister.

Pictured is the dedication of the Lutheran church on February 1, 1953. Worship services—ranging from Episcopal, Mormon, and Catholic to the interdenominational Grace Community Church—are part of Boulder City's culture. From the beginning of the town's history, churches played music through outside loudspeakers on Sunday mornings and evenings. "Amazing Grace" and other hymns pealed across the houses, and streets, and newly planted shrubs.

RESTAURANTS AND BUSINESSES

Camp Williston, also known as Camp Sibert, was established just before World War II to provide military police training and protection for dam facilities and local industries. After the US Army abandoned the post in 1944, the camp was put up for sale, and the buildings were eventually sold and moved. Pictured are a few former Camp Williston annex structures that were moved to Arizona Street in January 1947 for use as extension classrooms at the high school.

A view of the annex structures situated on the parking lot on Ave G.

CHAPTER 3

HOUSING

Houses and dormitories for Six Companies and Bureau of Reclamation employees appeared to rise from the desert overnight. The Bureau of Reclamation planned the town for 6,000 residents during the dam's construction period; however, 1,500 residents needed a place to live after the completion of the dam. Only those who had jobs on the project and their families were permitted to live within the federal reservation. This is an aerial view of what housing looked like on Avenue G and Avenue H in 1945.

Bureau of Reclamation housing is shown here on Colorado Street at Nevada Way in 1932. Houses located on the north side of Colorado Street were built between 1931 and 1932 for field and office engineers working on the Boulder Canyon Project. Robert E. Dolin was among the many employees. Dolin served as a machine shop foreman for the bureau after the dam was completed; he was among those who had worked for Six Companies as a machinist.

During a planning commission meeting held in 1931, a plan was adopted for the supervision of various functions. This was exercised in executing the Six Companies contract and included things such as construction; purchasing, housing and feeding workers on the job; transportation; insurance; and hospitalization. The building of the housing for dam workers was supervised by Felix Kahn, of Macdonald & Khan; K.K. Bechtel, of the W.A. Bechtel Company; and Frank T. Crowe, superintendent of construction for Six Companies.

HOUSING 63

One of the Six Companies homes located on Avenue B belonged to Bernard "Dick" or "Smokey" and Jessie Farr (pictured) in 1933. The Farrs resided in this home during the entirety of the construction of the dam. Once the dam was complete, the Farrs decided it was time to move on and sold the home to Congressman James G. Scrugham of Reno. Scrugham fell in love with Boulder City during his visits and wanted to enjoy the benefits of the growing community.

HOUSING

Most homes in the area were built of wood frames or brick, but four were constructed of rock, and three of them still stand today. One of the earliest rock homes (pictured) was built on Avenue G on an empty lot left after Six Companies tore down several blocks of worker housing.

HOUSING 65

The homes along Birch Street housed those who worked for Los Angeles Department of Water and Power (LADWP), also known as the Los Angeles Bureau of Power and Light. One of those homes belonged to pioneer resident William E. French, who worked for Six Companies during the construction days. After completion of the dam, he worked for the Bureau of Reclamation, Bureau of Mines, and Basic Magnesium Inc. plant in Henderson until he became an employee of the LADWP maintenance department. In 1958, he and his wife, Edna, purchased Uptown Hardware.

Babcock and Wilcox Company built apartment buildings, each of which could house up to four families, along the west side of Avenue B in 1932 and 1933. These were part of the Six Companies plan to build homes to house employees who were working on the dam or in the shop.

Housing 67

Firefighters show off their new Howe engine on Colorado Street on December 28, 1931. The driver of the truck is Robert Hewes, who was in charge of fire detail for the Bureau of Reclamation. In 1931, the need for motorized fire apparatus to replace hand-drawn pioneer equipment was recognized. Fire was a constant problem for the new town, and firefighters relied solely on two-wheel hand-drawn hose carts and chemical extinguishers. These were inspected semimonthly by the superintendent.

Pictured here are some of the homes that housed employees of the Bureau of Reclamation along Colorado Street in 1932. Of the homes pictured, one belonged to safety engineer Robert Cary, one to Paul Bonnell, and another to Joseph R. Drake. All of these homes were built between March 1931 and June 1932. Occupancy changed often due to the shifts in the types of engineers needed for various phases of the dam's construction.

HOUSING 69

J.R. Alexander lived in this home during the dam's construction days in the 1930s; it later belonged to Clarence Aryo. Alexander reported to Walker R. Young, who was the official in charge of the projects under construction in Boulder City. Alexander was one of the few members on Boulder City's first school board and also served on the Bureau of Reclamation District Council for Boulder City.

This 1932 photograph shows Bureau of Reclamation homes along the north side of Park Street. These were some of the grandest residences in Boulder City that the Bureau of Reclamation was awarded for contract and were in close proximity to the administration complex. One of these homes belonged to John and Fanny Connolly. Fanny was a Boulder City Hospital administrator for 13 years.

HOUSING

The house at the end of Park Street (center background) was home to P.S. "Jim" Webb. Webb was a construction contractor and built the first business and residential structures in Boulder City. He also formed the Boulder City Development Company, was a partner in Boulder Dam Tours, and organized the Boulder City Builders Supply Company, dealing in lumber and materials. His colorful career goes deep and is an example of how this town came to be.

HOUSING

William S. and Mae W. Miller bought this house on Avenue C when the dam's construction has been completed. The boy pictured here is their son Ernest A. Miller. The Miller family came to town in 1932, and until 1937, William and his brother Ernest P. Miller operated the meat concession at Manix & Vaughn department store. William was also employed at the visitors bureau and was a member of Boulder City Lodge No. 37. Mae belonged to the Desert Chapter of the Order of the Eastern Star and once served as treasurer.

Housing

This is one of the homes that was built along Denver Street for Bureau of Reclamation project managers and engineers; the south side was reserved for other employees of the bureau. The first house on the north side belonged to Walker R. Young, chief construction engineer and one of the most respected engineers of the 20th century. He was also responsible for selecting the site where housing was built for the dam workers.

For three years prior to 1938, Charles H. Cady resided in the Boulder Dam Hotel while he planned and built a large residence overlooking Lake Mead on Denver Street (pictured). Subsequently, he constructed another home alongside it. Cady took an active part in the community life of Boulder City, assisting materially with every civic enterprise. The welfare of Boulder was vital to him, and he contributed his services to the chamber and the Rotary Club.

HOUSING 75

Homes built by Southern California Edison on Cherry Street are pictured here on August 29, 1939. During the construction of the dam, Cherry and Date Streets were central locations for the Six Companies' Anderson Brothers Mess Hall. Feeding 5,000 men each day required a state-of-the-art factory equipped with a kitchen, a bakery, ice storage, and a dining hall that could accommodate 1,200 men per sitting. More than 6,000 meals were prepared and served daily.

Babcock and Wilcox was the second-largest contractor on the Boulder Canyon Project and employed around 300 workers. This is one of the single-family dwellings that was located in the area originally reserved for commercial use at that time. The company also had a plant and office located one mile from the dam site where steel penstocks were fabricated for the pressure outlet and power system.

HOUSING

The houses located on California Street, Colorado Street (pictured), and Nevada Way were part of 12 single-family dwellings built by Babcock and Wilcox for reserved commercial use. The homes were purchased by the Los Angeles Bureau of Power and Light in 1937; they were removed in 1991 to make way for the construction of the Boulder City post office.

Before the Boulder City post office was located in the spot were a Babcock and Wilcox home (pictured) once stood, James L. Finney opened the first post office in Boulder City on April 15, 1931. Finney described the 16-by-20-foot shack as such: "If it wasn't for the heavy safe holding it down, it would have probably blown away."

Housing

This view looks north on California Avenue from New Mexico Street in 1931. Six Companies was contractually obligated to build temporary housing. These were manufactured homes designed by a Southern California firm. The design was based on those used to house workers who built the Panama Canal. The floor plan was simple and consisted of a living room, a kitchen, a bathroom, and bedrooms.

The married couple housing is shown here under construction. No one could build houses or sell so much as a radish in the federal reservation without a federal permit. Eighty percent of the workers had to live, eat, and sleep on the reservation in Boulder City. The housing quality was dictated by the discovery that the better the workers' conditions were, the faster they dug the dam.

HOUSING

Pictured are Frank (left) and Warren Shelton at their first home on Avenue L. Pfc. Frank Shelton was stationed at Camp Bowie in Texas and was appointed as principal candidate for the US Naval Academy in 1944. He grew up to be a nuclear physicist and resided in Colorado with his family. Warren Shelton was one of the town's track and football stars in 1948 and grew up to be an electrical engineer, residing in Boulder City with his mother, Jessie Shelton.

82　　　　　　　　　　　　　　　　　　　　　　　　　　　　　　　　　　　　Housing

Pictured here on July 6, 1933, is a privately owned cottage on a leased lot like many others on the street. One of the homes that was moved to Avenue L was purchased by Sylvester Bevel. His home once stood about 300 feet south of the Safety First building and was one of the Standard Oil deliveryman's home prior to Six Companies moving in.

HOUSING

This residence on Avenue I belonged to Jack and Lillian Weiler, who moved to town in 1931. Lillian was a homemaker, and Jack worked for Six Companies as a high scaler. He later joined the ranger force for the Bureau of Reclamation, where he was promoted to captain in 1942 and police sergeant after the rangers were reclassified in 1952. After the town's incorporation, Jack stayed with the Boulder Canyon Project as a guard supervisor at the dam until he retired in 1966.

Avenue I was one of the streets that received sewer service as part of the residence district community disposal system. In October 1931, it was announced that the system was next to be completed and that it would tie in with present temporary disposal plants. Homes on New Mexico Street, as well as a few on Avenues B and C, would be part of the system, too.

HOUSING

This is the home of the Ernest Ambrose Courture family. It was formerly home to Wilbur W. Weede, Boulder City's chief landscape gardener, and his family. Weede hailed from Oregon and was put in charge of the beautification project for the town's citizens and the visitors who would be coming from near and far to visit the dam. The first order he placed was for 9,000 trees from Nevada and California nurseries.

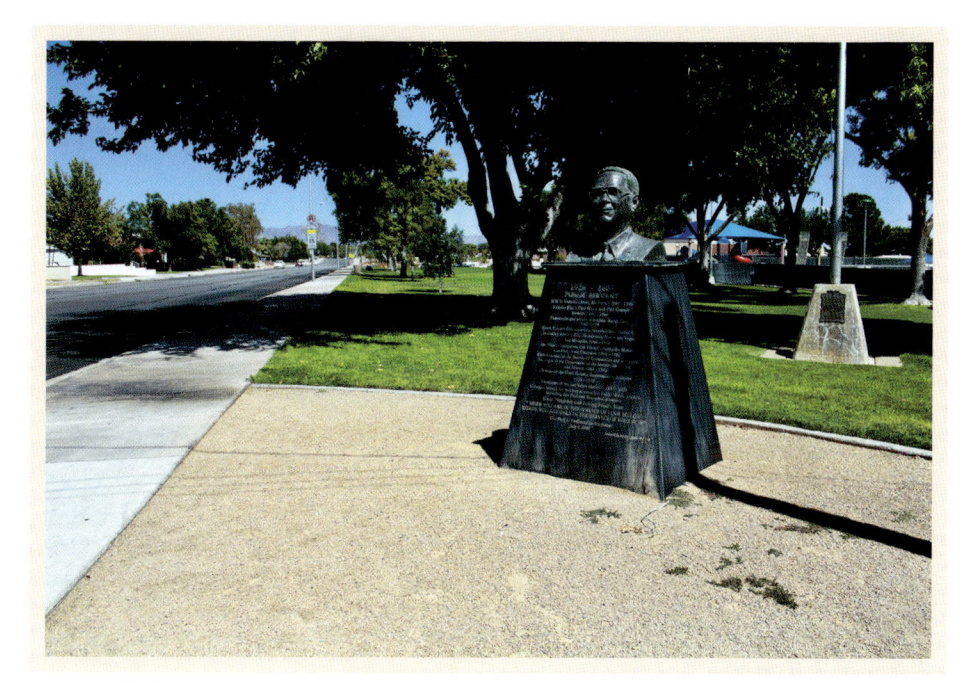

Demountable homes are shown along Fifth Street in the early 1940s. These were not permanent houses, and they could be taken down in sections and moved or sold. As the district expanded, so did the need to alleviate the housing shortage for the community during the wartime boom. Near the start of World War II, Thomas Buck Construction was contracted to build 100 little framed homes.

HOUSING 87

This Boulder City home, located on California Street, was completed by George David Stegmuller in 1938. "The only air conditioning was a homemade swamp cooler that consisted of excelsior with water dripping on it while a fan in front of it pulled air through the wet wood fibers." George was an employee of the dam during its construction, working his way up from being a water boy to becoming an oiler on the big steam shovels; he also witnessed one of the 96 accidental deaths that happened on the dam.

As a Boulder City pioneer, Rose Lawson lived on this avenue starting in 1931. She described Avenue F as follows: "In the beginning, this area was still fresh with sandy roads, and even though the homes were just built and had electricity, there was still no water yet. We would have to go to a railroad car, which brought water from Las Vegas."

HOUSING 89

One of the first settlements in Boulder City was McKeeversville, pictured here in 1937. The name came from a government cook, Michael McKeever, who lived here with his family. It was known as a squatters' camp where poorer families lived until better housing became available. Many locals called the area "the other side of the tracks," as the railroad tracks ran between McKeeversville and the Boulder City township. The remnants of the 1930s camp still stand today.

The family residence of the bureau's office engineer, John C. Page, is pictured here on June 1, 1932. One of his daughters—either Jean or Mildred—is shown in the photograph. Page began as a topographer surveying canal sites in Colorado in 1909 and became office engineer and chief administrative assistant for the Boulder Canyon Project in 1930. By 1935, he advanced to become the head of the engineering division. On January 25, 1937, he assumed the commissioner's office after filling in as acting commissioner the previous year.

Housing

This private home was built by Glover E. Ruckstell in the mid- to late 1930s. Ruckstell purchased three Ford Trimotor aircraft in 1931 and set up operations for Grand Canyon Airlines at Bullock Field in Boulder City to provide scenic flights over the Boulder Dam, Grand Canyon, and Death Valley. In 1933, he and his business partners built the Boulder Dam Hotel. On June 15, 1936, an agreement with Howard Hughes and Trans World Airlines (TWA) brought passenger service to town.

Marcus "Mark" and Jessie Shelton lived on Avenue L until Mark engaged a contractor to build a house on Avenue I (pictured). The couple came to Nevada in 1931, when Mark obtained a job at the dam, and they lived in one of the first six tents in McKeeversville. After Mark passed away following an auto accident, Jessie worked at the checking station at the west gate of the dam until 1945. She later became a local telephone operator.

Housing

Pictured is the Six Companies Executive Lodge in 1933. The executive lodge "guest house" was built to be used by Six Companies directors, their families, and guests. The lodge is 2,000 square feet and has six bedrooms, a living room, and a dining room with rare china set-up. The property is owned by the Los Angeles Department of Water and Power and is still used for executive meetings or historic tours.

In 1932, the Bureau of Reclamation set aside several blocks east of Boulder City where workers would be allowed to lease a lot from the government and build their own homes as long as they followed specifications outlined by Boulder City administrators. These blocks included Avenues K, L, and M (pictured). They would become part of the economic importance of the town's history and stand as historic reminders of the craft of the dam workers who built them.

Housing

DISCOVER THOUSANDS OF LOCAL HISTORY BOOKS FEATURING MILLIONS OF VINTAGE IMAGES

Arcadia Publishing, the leading local history publisher in the United States, is committed to making history accessible and meaningful through publishing books that celebrate and preserve the heritage of America's people and places.

Find more books like this at
www.arcadiapublishing.com

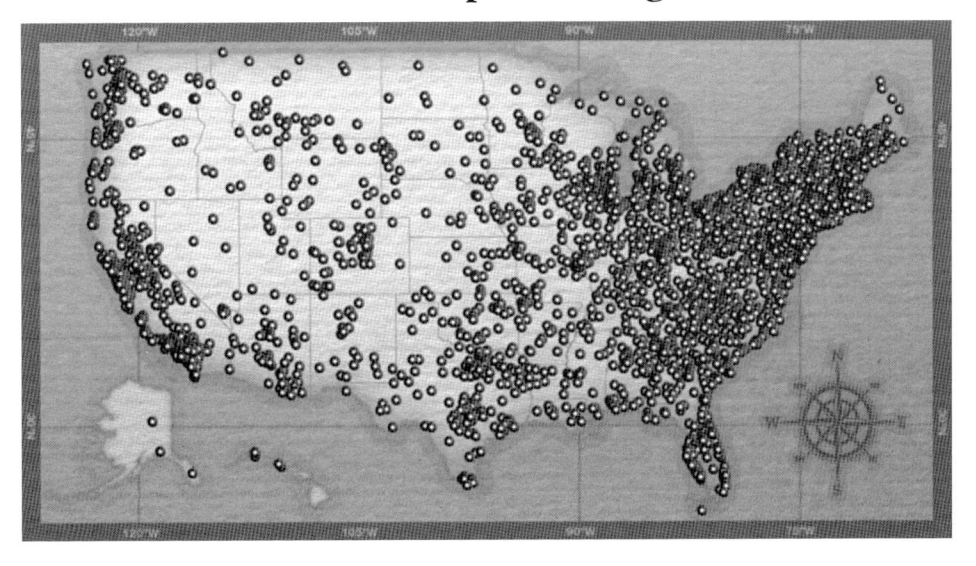

Search for your hometown history, your old stomping grounds, and even your favorite sports team.

Consistent with our mission to preserve history on a local level, this book was printed in South Carolina on American-made paper and manufactured entirely in the United States. Products carrying the accredited Forest Stewardship Council (FSC) label are printed on 100 percent FSC-certified paper.

MADE IN THE USA